Andean Condor
(Vultur gryphus)

The New World Vultures

There are seven species of New World vulture, including the two types of condor. They occur across the whole width of the Americas between the Pacific and the Atlantic, and from the latitude of the Great Lakes in the north to Cape Horn in the extreme south. The altitude of their habitat ranges from sea level to the highest mountains, and they occur in every type of vegetation. Some have acquired urban habits but the larger species learned long ago to shun towns and their inhabitants.

All vultures are scavengers, that is to say they feed on the carcasses of dead creatures.

It seems that the New World vultures evolved towards their present form at an earlier time than the Old World ones, and that the two groups do not necessarily share a common ancestry. They are an example of convergent evolution, whereby similar types have evolved and adapted for similar ways of living. Old World vultures derive from primeval types of eagle and hawk, while New World vultures once had more species than now and were until recently thought to descend from storks, with which they share behavioural links and certain internal and external characteristics. Ornithologists were long reluctant to admit this proposition although Thomas Huxley first advanced it in 1876. Recently the evidence for this theory has again been questioned, leading to a proposal to take these Cathartidae out of the order Falconiformes and put them into a new order of their own, Cathartiformes.

Some ancestral fossils in Argentina were a remarkable find. The earliest known condor, *Argentavis magnificens* from the Pleistocene, had a wingspan of 7 metres. By comparison, today's Andean Condors have a spread of up to 3.2 metres.

19th century observers such as Humboldt probably exaggerated with accounts of 4.3 metres for some condors he saw at a distance in Ecuador. The relatively slender Wandering Albatross has a wider wingspan, but Andean Condors remain the world's largest flying birds.

The New World Vultures
by
Nigel Hughes

Class AVES
Order FALCONIFORMES
Suborder CATHARTAE
Family CATHARTIDAE (New World Vultures)

Distribution map

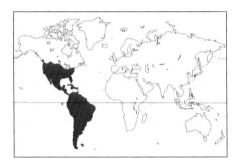

Andean Condor *(Vultur gryphus)*
California Condor *(Gymnogyps californianus)*
King Vulture *(Sarcoramphus papa)*
American Black Vulture *(Coragyps atratus)*
Turkey Vulture *(Cathartes aura)*
Greater Yellow-headed Vulture *(Cathartes melambrotus)*
Lesser Yellow-headed Vulture *(Cathartes burrovianus)*

Text and illustrations by Nigel Hughes FLS FRGS

I can think of no bird family that has so negative a reputation, nor such a false one as the New World vultures. They had to bear this burden of unpopularity, ignorance and prejudice during the 19th and 20th centuries, and to this very day in many places. Part of the trouble is a mistaken attitude among livestock farmers who believe that vultures of all types kill cattle and spread disease. It is true that some of these birds will occasionally tip a dying animal over the edge of a precipice and then go down to eat its corpse, but it is very unusual, especially as their claws are not adapted for attack. It has been said that New World vultures kill and eat newborn calves and lambs, but the evidence is scant. People who earned their living from guano production also thought their interests were damaged by condors eating eggs and chicks of the sea birds that provide this agricultural fertilizer on the Pacific coasts of Peru and Chile. Again, the damage is negligible in proportion to the numbers involved. And, as for diseases, vultures reduce them, as I will explain later.

Vultures have been poisoned intentionally with strychnine and unintentionally with lead. Lead bullets part into small fragments on impact inside hunted animals: the least amount of lead eaten by a vulture from an abandoned carcass will often kill the bird and always damage it in the short or long term by its toxicity, as may other metallic débris. This too is mentioned again later.

I should like to quote two great naturalists of former centuries, because their unfortunate judgements have supported some damaging attitudes towards these birds, after thousands of years during which both Old World and New World vultures were respected, even revered, and for good practical reasons. The main one of these is that the natural world would be in a sorry state without them.

Count Buffon, the famous French zoologist who lived through most of the 18th century, wrote:

"People have given eagles the first rank among birds of prey, not because they are stronger and larger than vultures but because they are more generous, that is to say less meanly cruel. Eagles' ways are prouder, their attitudes braver, their courage nobler... Vultures on the other hand are nothing more than basely greedy. They eat living creatures only when they cannot find dead ones..."

Paul Géraudet, despite being a most accomplished and helpful ornithologist, nevertheless wrote in the 1950s of vultures 'ignobly nourishing themselves on dead bodies and filth'. It's the 'ignobly' that worries one most. Other commentators join in: 'Filthy', 'disgusting', 'indolent', 'rapacious', 'smelly'… No end of foolish and misguided insults.

The scavenging nature of the New World vultures developed from an original predatory character involving living prey. In their natural territories there is always more live prey than dead, except in times of epidemic disease. These would-be scavengers will have lost a lot by abandoning their capacity to hunt live animals, but the change worked to their advantage when they developed much larger wings, making it possible to soar efficiently and economically. All species flap their wings as little as possible, although the American Black Vulture is more active than the others in that way. The condors, above all, can cover hundreds of miles in a day, sustaining their soaring flight by an intelligent use of upward air currents. These they find above bare sunny ground and at the windward edges of cliffs and escarpments. No more high-energy, high-speed hunting for them. One consequence is that vultures are less aggressive than other birds of prey, especially between themselves. The Black American Vulture is exceptional in being more inclined to violent quarrelling.

Their careful husbanding of energy is helped by having no natural predators to contend with, although occasionally large felines, raccoons and snakes take eggs and chicks. However, there *is* a vulnerable time for all vultures, which is when they are so replete with food that they cannot take off as promptly as they would like. On the rare occasions when one is attacked in this situation it can look after itself by vomiting in its enemy's face – quite an effective deterrent.

Some characteristics are common to all or nearly all the seven species:

They all have a remarkable acuity of vision. The lesser Yellow-headed Vulture, for instance, can spot the dead body of a small rodent on the ground from 3,000 metres above. The two types of condor, interested in larger carcasses, can spot prey from heights greater yet. However,

California Condor
(Gymnogyps californianus)

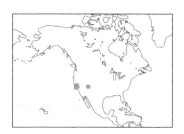

higher-flying birds are not always the originators of a downward flight on to food. They all keep watch for companions leaving an evenly spaced grid of observant birds to descend on a carcass: vultures from a wide circle will converge rapidly on the space left by the first one to go down, and all follow like water going down the plug-hole of a bath.

There used to be much debate about New World vultures' sense of smell. John James Audubon, naturalist and artist, was convinced that the Turkey Vultures of the Great Plains of the United States couldn't smell anything. To prove his point he set out some putrid meat, having chosen it for its appalling stench. The vultures wouldn't come near it, but this was because they could smell it from miles away and didn't like it.

However, Audubon was partially right. The Cathartes species (Turkey, Greater Yellow-headed and Lesser Yellow-headed Vultures) do have an acute sense of smell, while the other four New World species (two condors, King Vulture and American Black Vulture) have none.

Since one would have thought a sense of smell useful to every species, one has to wonder why this is not the case with these. The answer must be that their excellent vision allows them to manage without, and that the surprisingly large amount of energy required by the brain for processing the sense of smell is better used for other purposes, such as flying and dismembering carcasses: the brain uses more energy in proportion to its weight than any other organ in the body.

Perhaps because they are not territorial and therefore have no reason to complain loudly to their fellows about encroachment, New World vultures have no syrinx, the vocal organ of birds. They cannot cry out or sing, only hiss or grunt or sneeze.

All New World vultures have bare heads and necks, enabling them to keep themselves clean after putting their heads inside carcasses. They are indeed fastidious creatures, clean by nature and particular about the healthy and orderly state of their plumage. Another property of those bare areas is that they can be made to colour up strongly, thus increasing the local blood supply and getting rid of surplus body heat.

This is particularly important with birds that have been soaring in cold air at great altitudes, keeping themselves warm up there by retracting their heads and necks into their neck feathers and tucking their feet into the plumage below their tails. On coming down to the ground, they can find themselves in very hot conditions of bright sunlight and still air – a sudden change. Quick adaptation and avoidance of heat stress are essential. Not only do their bare heads and necks help regulate their temperature by radiating heat; there is also the phenomenon of urohydrosis. This is the vultures' habit of excreting on to their feet, which refreshes them wonderfully. The urine element evaporates through the chalky element of their waste, thus cooling the birds in the manner of those pioneering fridges made of chalk. One poured water into a depression at the top and the chest became cold inside. If one didn't know about urohydrosis, one would wonder why the birds' feet appeared to be whitewashed. Another advantage of the phenomenon is that it disinfects the surfaces of the feet, which get mucky when the bird is dismembering what it has removed from a carcass. Sunlight bakes the bacteria.

Most birds have a more complex breathing apparatus than do mammals. Vultures, which soar at a great height then come to ground level according to their requirements, need this complexity more than most. How is it that they can adapt so well to altitudes at which humans could not function successfully without oxygen apparatus? And how can vultures ascend to those heights without the periods of acclimatization adopted by humans climbing even moderately high mountains?

There are three means by which these things are achieved. One is sufficient storage of oxygen-charged air, especially when that air is thin with altitude. Another is the respiratory system, and the third is the deployment of different types of haemoglobin in the blood.

A flying bird consumes more oxygen than a running mammal of the same size, yet it is never seen to be out of breath, due to the total volume of the bird's respiratory system being about three times that of a mammal of corresponding size. The bird has a number of air reservoirs – air sacs, lungs and storage spaces within the major bones. Birds have several pulmonary receptacles corresponding to, and proportionately more capacious than, the two lungs of mammals. We need concern ourselves only with six: two anterior air sacs, two lungs and two posterior air sacs.

When the bird inhales, the posterior air sacs are filled with fresh air and the lungs empty their used air into the anterior sacs, which thereby become full. On the out-breath, the lungs are filled with fresh air from the posterior air sacs, and the used air in the anterior sacs leaves the bird via the throat. On the next inward breath, the lungs expel their by then exhausted air into the anterior sacs, while the next lot of fresh air arrives in the posterior sacs. And so on. It will be seen that it takes two complete breaths to deal with one complete intake of fresh air.

The posterior air sacs are also connected to air reservoirs in the major bones.

The last of these three ingenious contrivances is the variety of haemoglobins found on the surface of vultures' red blood cells. These absorb oxygen into the blood. Mammals have only one sort, but many birds have two, a high-altitude and a low-altitude form. The type is switched according to the blood's demand for oxygen. The highest-flying vulture of all is an Old World species, Rüppell's Vulture, observed at 11,300 metres, roughly the height of Mount Everest, above sea level over Abidjan in West Africa. It has four different types of haemoglobin.

Vultures' bills are sharply hooked at the front and have serrations all along their convergent edges, for cutting meat away from carcasses. On the upper part of the bill are the nostrils, through which one can see from side to side. This feature makes it as easy as possible for the birds to blow out anything that may have got stuck there while foraging.

These birds' feet are variously but not emphatically webbed. Their claws are more in the nature of toe-nails than talons: they haven't the needle-like character of hawks', and are therefore neither developed nor fitted to seize moving prey. They have fewer scales on their shanks than most birds, the texture on most of their feet being like that of

King Vulture
(Sarcoramphus papa)

shark-skin with its small convex papillae. The halluxes or hind toes of all New World vultures are vestigial: they have little strength and, being set higher than the forward three toes on each foot, are not much use in perching. The only purpose I can deduce is that they must be a help in cleaning and arranging their plumage.

This brings us back to general hygiene. As already mentioned vultures keep their outsides extremely clean, bathing in fresh water whenever they can. But a greater wonder is going on inside. In spite of the stricken state of the animals they ingest, including those that have died from infectious diseases, the vultures continue in perfect health, thanks to the potency of their gastric juices. As well as disintegrating bones (in different degrees according to species) they neutralize toxins and the aggressive pathogens that are present in carcasses. These include the bacteria of botulism, anthrax and black-quarter (for example). It seems that certain viruses are also destroyed, perhaps even ebola, the disease that surged up again in Africa to such dire effect in 2014. Disaffected farmers should no longer be able to claim that vultures have a negative effect with regard to diseases since, in effect, they eat them. Admittedly, not all the bacteria of diseases such as anthrax and black-quarter are eliminated but vultures certainly curtail them. The pathogens can survive for years in the soil below a carcass, but they would do so to a far greater extent if there were no vultures to dismantle the dead animals, which they generally do in a very short time from the animal's death – a very positive factor.

Vultures are not, and cannot be, early risers (you can imagine the 19th century's anthropocentric response – 'lazy', 'indolent', 'apathetic'). They must wait until the sun has become high enough to start the upward air currents upon which they can ascend. While they wait they spread their wings, warm themselves as with solar panels and arrange their flight feathers for the day. A soaring bird's primary feathers take on an upward bend during the course of a day's flying, but they straighten back to their proper shape when the bird is resting with its wings spread in bright sunshine.

If the weather is not right for flying, they have to remain grounded, motionless but awake, while they wait for better times. The contents of their capacious crops (enlarged parts of the oesophagi used to store food temporarily) enable them to sit out bad weather for many days. An Andean Condor can consume and store within its digestive system two kilograms in one meal. Charles Darwin recorded one of them fasting for four or five weeks. Turkey Vultures can certainly do without food or water for two weeks. The birds can also lower their metabolic rate in times of inactivity.

The crops have a social function: if its crop is distended with food and also brightly coloured, a vulture cannot hide that fact from hungrier companions. Those that are not so full can then, one supposes, assert a feeling of entitlement and crowd in to share whatever is available.

American Black Vultures and Turkey Vultures, and sometimes the condors, roost in groups; other species only as individuals or pairs.

All the New World vultures appear to mate for life. They do not go in for formal nests at breeding time, preferring a meagre assemblage of twigs and grass on the ground, or on a rocky shelf, or the mouth of a cave in a cliff. The smaller types may use the top of a broken tree trunk. The better-studied species seem to have no system of communal vigilance. The birds are secretive about leaving or approaching their eggs. Another defence is to make sure the chosen site smells absolutely foul so that possible predators may suppose that no good can come of trying to eat either eggs or chicks. Nevertheless there have been cases of a huge California Condor being deprived by ravens of its single egg.

New World vultures do not migrate, except for two subspecies of the Turkey Vulture and some of the American Black Vultures, both to be described later.

American Black Vulture
(*Coragyps atratus*)

Species Notes

The **Andean Condor** (*Vultur gryphus*) has a wingspan of 3.3 metres and is the largest flying bird the world. Its adult weight is up to 15 kilograms. It is the only dimorphic species (i.e. visibly different between the sexes) among the New World vultures, the female's head being greyer and pinker and lacking the male's prominent fleshy crest.

Both this and the California condor reach breeding condition at 6 years and lay one successful egg every two years thereafter. If the first egg is lost, they can replace it with a second one. It takes more than a year for condors to bring their chicks to independence, including an incubation period of about 55 days, which accounts for the biennial laying of eggs. The slow rate of reproduction is compensated by longevity. They can live, it seems, for more than seventy years, although not so long in the wild.

Andean Condors feed on the rocky coasts of the Pacific and on beaches where they can find all sorts of dead creatures washed up. In the 18th century they were a common sight round Lima and its port of Callao. At the other extreme of the altitude scale they search the snow fields of the high Andes for mammal casualties, such as guanacos killed by avalanches. Not so very long ago there were unusual sightings of an Andean Condor near Santos on the Atlantic coast of Brazil, far from its usual habitat at that latitude.

Condors may start their day roosting in a deep canyon waiting for mid-morning updraughts to lift them to the level of a plateau above, where they can bask in the sun before finding further thermals to take them even higher. Then they can soar away to great distances, on the look-out for food all the way. They may travel hundreds of miles in a day, although not so far as was reckoned by the 18th century ornithologist and would-be aviator Santiago Cardenas. He studied the condors outside Lima, observing them from the summit of the Cerro San Cristóbal and, not having a watch, judging their speeds by his own out-of-breath pulse. He had them flying at 600 miles an hour and was so pleased by his findings that he devised a condor-based flying machine with which he planned to operate an air mail service between Lima and Buenos Aires. His hopes appear to have had the encouragement of the viceroy but sadly they came to nothing.

Turkey Vulture
(*Cathartes aura*)

The **California Condor** (*Gymnogyps californianus*) was once common from Mexico to well north into Canada, and from the shores of the Pacific to beyond the Mississippi, Missouri and Ohio rivers. However, in the latter half of the 20th century it became clear that this huge bird was almost extinct. In 1986 there were only 22 individuals left. The last survivors were caught the following year and nearly all were induced to breed in captivity. Now there are just over 400 individuals all told. Despite the usual setbacks of such a project there is reason to hope that they will recover in at least some parts of their former range. But one must emphasise that wild-born offspring are still few in number and that the species remains in danger of final extinction. Present release areas include the Coastal Range in California, Big Sur on the Pacific coast, the Grand Canyon in Colorado, and Baja California in Mexico. Deaths are still occurring from lead poisoning (see below), accidental poisoning from other materials and collisions with power lines.

The California Condor has a wing span a third of a metre less than the Andean Condor's and does not have the latter's head ornament. In other respects the two species are similar. Alas, the California version is attracted by bright and pretty things, leading them to ingest all sorts of rubbish such as broken glass, bottle tops and those little aluminium tabs from drink cans. Any of these can be mortal, and so can lead bullets used in hunting: the condors eat from abandoned carcasses containing fragments of lead and are killed by it.

The **King Vulture** (*Sarcoramphus papa*) has been described as 'a flying can-opener'. Other species stand back from this lordly creature and wait for it to open tough old carcasses and break tendons, which its strength and size enable it to do. King Vultures do not usually assemble in great numbers at a carcass, a pair being more normal. They fly below other species in the search for food and are thereby well placed to be among the first to arrive. They are the most colourful of all the American vultures and in consequence are well represented in ancient indigenous art. In recent times they have abandoned heavily disturbed areas of forest for want of suitable carrion.

Their wingspan is nearly 2 metres. Each adult pair produces one egg per year. Nesting has not been much studied, the species being secretive in that direction. They are known to live for 30 years in captivity, presumably fewer in the wild.

The **American Black Vulture** (*Coragyps atratus*) has some notable differences from the other 6 species. It is more sociable among its own species, but more truculent at feeding sites. It has a magpie-ish liking for bright objects with which to decorate its otherwise minimal nest. They lay 2 eggs a year. They live to an age of around 16 in the wild, over 20 in captivity. Their wingspan is 1.4 metres. The northernmost populations migrate a short way southward in hard winter periods, then back again in the spring. They cannot survive in a snow-covered landscape because they cannot see their food.

It is the only species whose colouration is entirely black, grey or white. It has a longer neck than the other smaller species, and thereby can get its head further inside the rib-cage of a carcass. Also, as with other species, the large area of bare skin helps when the creature wants to lose heat. The Black Vulture tends to roost communally, sometimes together with Turkey Vultures. A benefit of this arrangement is that birds that have had an unsuccessful day can on the following day accompany those that have done better.

The American Black Vulture is the species whose flight involves the most flapping. I believe this is the only New World vulture species known to prey on live calves, but even that is a rare occurrence and such live prey tend to be ailing or weak.

This and the three Cathartes species (see below) breed at three years and lay two eggs a year. Incubation lasts around 40 days, and their offsprings' period of dependence is shorter than that of the condors and King Vulture.

The **Turkey Vulture** (*Cathartes aura*) is the most widespread vulture in the New World, occurring from Canada to Tierra del Fuego. Two of its four subspecies migrate in great numbers between the north of their range (see yellow area on species map) near the Canada-U.S.A. border, and the northern republics of South America. I am told that the passage of hundreds of them flying in groups over the isthmus of Panama is an astonishing sight. No American vulture likes to take short cuts over the sea because the lack of thermal up-draughts and consequent need to flap their wings is exhausting for them.

Apart from its usefulness as a scavenger, the Turkey Vulture has become useful in a new way. With its highly developed sense of smell it interests itself in the odour of commercial gas. The gas gives the vultures the impression of mildly deteriorated meat, and leads them to fly above leaks in pipelines. The birds can be seen circling in the sky above the place and thus they save a lot of costly searching.

The Turkey Vulture and the American Black Vulture are much the same size.They are the most numerous at feeding sites, where both species are often found together. Humboldt reported 70 to 80 birds feeding at one carcass; these days one can see far more than that on the refuse dumps of towns.

The **Greater Yellow-headed Vulture** (*Cathartes melambrotus*) was not scientifically described until 1964, despite being widespread in forested areas of Amazonic South America. One reason for so little being known about their ways is that they have learnt to keep clear of humans. On the whole, they fly higher above the forest than the other Cathartes and come less often into the open. They have wonderful vision and a keen sense of smell. Their wingspan is 1.8 metres. They produce 2 eggs each season. Their maximum age in the wild is not accurately known, but the only captive specimen in the world that I could hear of (and have a look at) is over 62 years of age. She lives in a zoo enclosure of tropical temperature in the company of a flock of pelicans.

The **Lesser Yellow-headed Vulture** (*Cathartes burrovianus*), wingspan 1.6 metres, 2 eggs a year, is less wary of mankind than the Greater Yellow-headed Vulture (*C. melambrotus*), and has a much more extensive territory. This is bounded to the north by the Mexican state of Vera Cruz and to the south by the basin of the River Plate. It prefers a more open terrain than that of *C. melambrotus* and can, with luck, be seen hunting over marshland near ports like Iquitos in Peru. They are closely related to *C. melambrotus* and the two species share most of their characteristics.

Greater Yellow-headed Vulture
(*Cathartes melambrotus*)

Threats and Hopes

I said earlier on that these creatures have no natural predators, but I'm afraid they have to contend with humans, who have done for them in a big way. Considering their immunity from dangerous bacteria in carcasses, vultures are surprisingly easily harmed by ingestion of chemical poisons and non-ferrous metals. Lead bullets found in the carcasses of shot game are another great menace. As I have mentioned, and will continue to do until the law about lead ammunition is universal, bullets disintegrate in hunted animals' bodies. Fragments of the metal are unknowingly picked up by vultures in meat left behind by hunters. The result is death or long-term lead poisoning. Where death is not immediate, lead and its harmful effects accumulate in the body, causing damage to various organs including those involved in fertility.

Poison, particularly strychnine, has been used for a long time in the Americas to get rid of mammals and birds that annoy farmers. It is an example of poison persisting in the food chain. Once a coyote, for instance, has died from the poison, a vulture will eat the animal and be killed by the same dose. The shooting of vultures is, I hope, on the decrease: word is getting around that it is unnecessary and undesirable. There are fines for doing it in some places and for some species. Unpleasant local customs involving Andean Condors (tying them to bulls, for example, in some obscure and not very ancient rite) still persist, but I think they too are going out of fashion.

To show what a serious impact can come from an unforeseen mistake, I will tell a tale of Old World vultures (with which we are not otherwise dealing). Aged cattle in India were being treated by their well-intentioned owners for the pains of arthritis, using an anti-inflammatory medicine called Diclofenac. Devout Hindus do not slaughter their cattle: they allow them to die from natural causes. The traditional next step is the elimination of their bodies by vultures. It will be remembered that the Towers of Silence, which epitomise a sensible approach to tropical hygiene, depended on vultures for their fulfilment. But alas, if a vulture eats any part of a Diclofenac-treated carcass, that bird dies more or less immediately or is seriously ill. Thus occurred a 90% reduction in the

Lesser Yellow-headed Vulture
(*Cathartes burrovianus*)

vulture populations of Nepal, with similar losses in parts of India and Pakistan. The next thing to happen was that, for want of vultures to control the public health situation, dogs and other mammal scavengers moved in and multiplied alarmingly, leading to a huge increase in the incidence of rabies. Since then humans have been dying from rabies at a rate of 30,000 per year, nearly three-quarters of them children. Diclofenac is now banned from veterinary use and, luckily for the world, vulture populations are slowly recovering.

Countries with vultures have started to realize that their predecessors in the Spanish colonies of America were right to protect these birds, including condors. The United States of America now lead the way: damaging or killing the migrant species (American Black and Turkey Vultures) can lead to a fine of up to $15,000, with an option for the court to impose a spell in jail. There is a vastly greater maximum fine in respect of the only other North American vulture species, the California Condor. Criminal attacks on the latter are met with a maximum fine of $100,000 plus a possible prison sentence. If the crime is corporate, up to $200,000.

The California Condor rescue programme has brought help from various sources. A salient one is California Assembly Bill 711 which comprehensively outlaws lead bullets used in hunting. But so far California is the only state to have enacted such a law. One can but hope that before long identical legislation will cover the whole of the USA and Canada and extend eventually to all countries that would benefit from it.

The brightest hope in all this is found in primary education. Children are almost invariably interested in conservation and the natural world. They are anxious to help and anxious to talk to all and sundry, including their parents. Teaching children natural history and conservation matters has now become commonplace in primary schools. Of course it is far from being universal, but it is helped along by a growing ecological awareness in many parts of the world. My own limited experience of it concerns Prince Albert's Blue-billed Curassow and the White-winged Guan, in Colombia and Peru respectively. The enthusiasm of teachers and children for the welfare of their natural surroundings has spread through their communities, thereby putting two precariously placed species into much safer situations.

It is interesting how in the Spanish American colonies of earlier times people knew very well what an important function their vultures were fulfilling. In 18th century Lima for instance you could be punished heavily for injuring one. Indeed, in certain countries of the Old World in ages long past you could be in even greater trouble for such a crime.

A decline of these remarkable creatures, and in some cases the extinction of a species, is an outcome that it is well within our power to change. But it will require both alertness and determination.